BODIES OF WATER

OCEANS

A Crabtree Roots Book

DOUGLAS BENDER

Crabtree Publishing
crabtreebooks.com

School-to-Home Support for Caregivers and Teachers

This book helps children grow by letting them practice reading. Here are a few guiding questions to help the reader with building his or her comprehension skills. Possible answers appear here in red.

Before Reading:

- What do I think this book is about?
    - *I think this book is about oceans and what they look like.*
    - *I think this book is about water and oceans.*

- What do I want to learn about this topic?
    - *I want to learn what oceans look like.*
    - *I want to learn how big oceans are.*

During Reading:

- I wonder why...
    - *I wonder why all oceans have salt water.*
    - *I wonder why oceans can have big or little waves.*

- What have I learned so far?
    - *I have learned that oceans can have ice.*
    - *I have learned that islands can be found in oceans.*

After Reading:

- What details did I learn about this topic?
    - *I have learned that ocean water is blue.*
    - *I have learned that all oceans are really big bodies of water.*

- Read the book again and look for the vocabulary words.
    - *I see the word* ***islands*** *on page 8 and the word* ***waves*** *on page 12. The other vocabulary words are found on page 14.*

This is an **ocean.**

All oceans are big.

Many **animals** live in the ocean.

You can find **islands** in most oceans.

Some oceans
have **ice**.

All oceans can have big and little **waves**!

Word List

Sight Words

all
an
and
are
big
can
find
have
in
is
little
live
many
some
the
this
you

Words to Know

animals

ice

islands

ocean

waves

33 Words

This is an **ocean**.

All oceans are big.

Many **animals** live in
the ocean.

You can find **islands**
in most oceans.

Some oceans have **ice**.

All oceans can have
big and little **waves**!

BODIES OF WATER

OCEANS

Crabtree Publishing

crabtreebooks.com 800-387-7650

In Canada: We acknowledge the financial support of the Government of Canada through the Canada Book Fund for our publishing activities.

Hardcover	978-1-4271-5592-4
Paperback	978-1-4271-5598-6
Ebook (pdf)	978-1-4271-3381-6
Epub	978-1-4271-3441-7
Read-along	978-1-4271-5616-7
Audio book	978-1-4271-5622-8

Library and Archives Canada Cataloguing in Publication
Title: Oceans / Douglas Bender.
Names: Bender, Douglas, 1992- author.
Description: Series statement: Bodies of water | "A Crabtree roots book".
Identifiers: Canadiana (print) 20210190566 | Canadiana (ebook) 20210190582 | ISBN 9781427155924 (hardcover) | ISBN 9781427155986 (softcover) | ISBN 9781427133816 (HTML) | ISBN 9781427134417 (EPUB) | ISBN 9781427156167 (read-along ebook)
Subjects: LCSH: Ocean—Juvenile literature. | LCSH: Oceanography—Juvenile literature.
Classification: LCC GC21.5 .B46 2022 | DDC j551.46—dc23

Published in Canada
Crabtree Publishing
616 Welland Avenue
St. Catharines, Ontario
L2M 5V6

Published in the United States
Crabtree Publishing
347 Fifth Avenue
Suite 1402-145
New York, NY 10016

Written by: Douglas Bender
Designed by: Rhea Wallace
Series Development: James Earley
Proofreader: Janine Deschenes
Educational Consultant: Marie Lemke M.Ed.
Photographs: Shutterstock: Vladimir3d: cover, p. 1; Svetlana Turchenick: p. 3, 14; Katerina Klio: p. 5; stockphoto-graf: p. 6, 14; BlueOrange Studio: p. 8-9, 14; INTERTOURIST: p. 11, 14; Four Oaks: p. 13, 14

Printed in Canada/072023/CPC20230717

Library of Congress Cataloging-in-Publication Data
Names: Bender, Douglas, 1992- author.
Title: Oceans / Douglas Bender.
Description: New York, NY : Crabtree Publishing Company, [2022] | Series: Bodies of water - a Crabtree roots book | Includes index.
Identifiers: LCCN 2021017110 (print) | LCCN 2021017111 (ebook) | ISBN 9781427155924 (hardcover) | ISBN 9781427155986 (paperback) | ISBN 9781427133816 (ebook) | ISBN 9781427134417 (epub) | ISBN 9781427156167
Subjects: LCSH: Ocean--Juvenile literature.
Classification: LCC GC21.5 .B45 2022 (print) | LCC GC21.5 (ebook) | DDC 551.46--dc23
LC record available at https://lccn.loc.gov/2021017110
LC ebook record available at https://lccn.loc.gov/2021017111